DE

SECRETS
OF THE
ANIMAL WORLD

JELLYFISH
Animals with a Deadly Touch

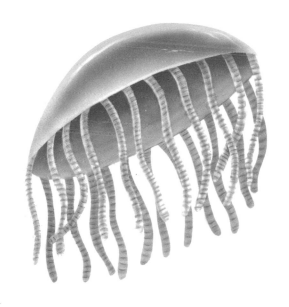

by Eulalia García
Illustrated by Gabriel Casadevall and Ali Garousi

Gareth Stevens Publishing
MILWAUKEE

For a free color catalog describing Gareth Stevens' list of high-quality books and
multimedia programs, call 1-800-542-2595 (USA) or 1-800-461-9120 (Canada).
Gareth Stevens Publishing's Fax: (414) 225-0377.
See our catalog, too, on the World Wide Web: http://gsinc.com

The editor would like to extend special thanks to Jan W. Rafert, Curator of Primates
and Small Mammals, Milwaukee County Zoo, Milwaukee, Wisconsin, for his kind
and professional help with the information in this book.

Library of Congress Cataloging-in-Publication Data

García, Eulalia.
    [Medusa. English]
    Jellyfish: animals with a deadly touch / by Eulalia García; illustrated by Gabriel Casadevall
and Ali Garousi.
        p. cm. — (Secrets of the animal world)
    Includes bibliographical references and index.
    Summary: Describes the physical characteristics, habitat, behavior, and life cycle of this
umbrella-shaped sea animal.
    ISBN 0-8368-1648-X (lib. bdg.)
    1. Jellyfish—Juvenile literature. [1. Jellyfish.] I. Casadevall, Gabriel, ill. II. Garousi,
Ali, ill. III. Title. IV. Series.
    QL377.S4G3713   1997
    593.5'3—dc21   96-45107

This North American edition first published in 1997 by
Gareth Stevens Publishing
1555 North RiverCenter Drive, Suite 201
Milwaukee, Wisconsin 53212 USA

This U.S. edition © 1997 by Gareth Stevens, Inc. Created with original © 1993
Ediciones Este, S.A., Barcelona, Spain. Additional end matter © 1997 by Gareth
Stevens, Inc.

Series editor: Patricia Lantier-Sampon
Editorial assistants: Diane Laska, Rita Reitci

Printed in the United States of America

1 2 3 4 5 6 7 8 9 01 00 99 98 97

CONTENTS

THE WORLD OF THE JELLYFISH

Where jellyfish live

Jellyfish, together with other members of the scientific phylum Cnidaria (nee-DAR-ia), make up one of the most common groups of sea animals. Jellyfish can be found in all the world's oceans, from warm tropical seas to the coldest seas, at depths between 3,280 feet (1,000 meters) and shallow coastal waters. A few species of jellyfish, however, also live in freshwater lakes, rivers, and ponds.

Many jellyfish are carried by strong currents to beaches and other places along the seashore. This is especially true in late summer in some areas. The jellyfish's ability to sting can cause

Jellyfish are transparent, elegant, mysterious creatures.

discomfort to swimmers. Beaches sometimes must be closed to swimming because of jellyfish.

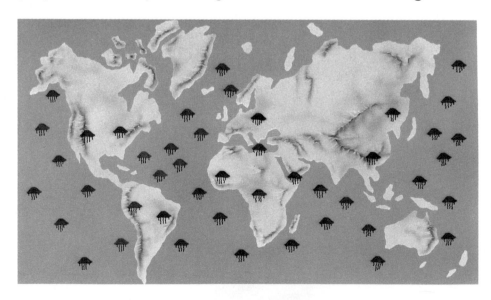

Jellyfish live in all Earth's seas. Some species also live in freshwater habitats.

Sea nettles

Despite their delicate appearance, jellyfish hide a lethal weapon. The slightest contact with a jellyfish tentacle can be painful, even fatal. This is why they have the name *cnidaria*, which means "nettle." Cnidarians appear in two forms: the polyp and the jellyfish. The polyp is sessile; a stalk attaches it to seaweed or other objects. The polyp reproduces asexually. The jellyfish is free-swimming and reproduces sexually.

Some jellyfish live in colonies along with polyps. But most jellyfish are solitary. They measure between a few inches (cm) and 11 feet (3.5 m) in diameter.

This jellyfish may seem to be eating algae, but jellyfish are carnivores. They eat fish and shellfish.

Cnidarians use their tentacles to capture prey. The tentacles have cells that sting.

Jellyfish and polyps

There are 9,000 species of cnidarians, which can be divided into three types, or classes: hydrozoa, which include polyps and jellyfish; scyphozoa, which include the true jellyfish; and anthozoa, which include anemones and coral.

The sea tomato, a common anemone on coastlines, is a bright red polyp with a crown of tentacles that it hides when threatened or when the tide goes out. The "strawberry" variety of sea tomato is red with green spots.

The Portuguese man-of-war is a colony of polyps with a large buoy — a modified jellyfish. The polyps and tentacles, which can measure

Cnidarians can be divided into three types. The most beautiful and spectacular jellyfish belong to the scyphozoa.

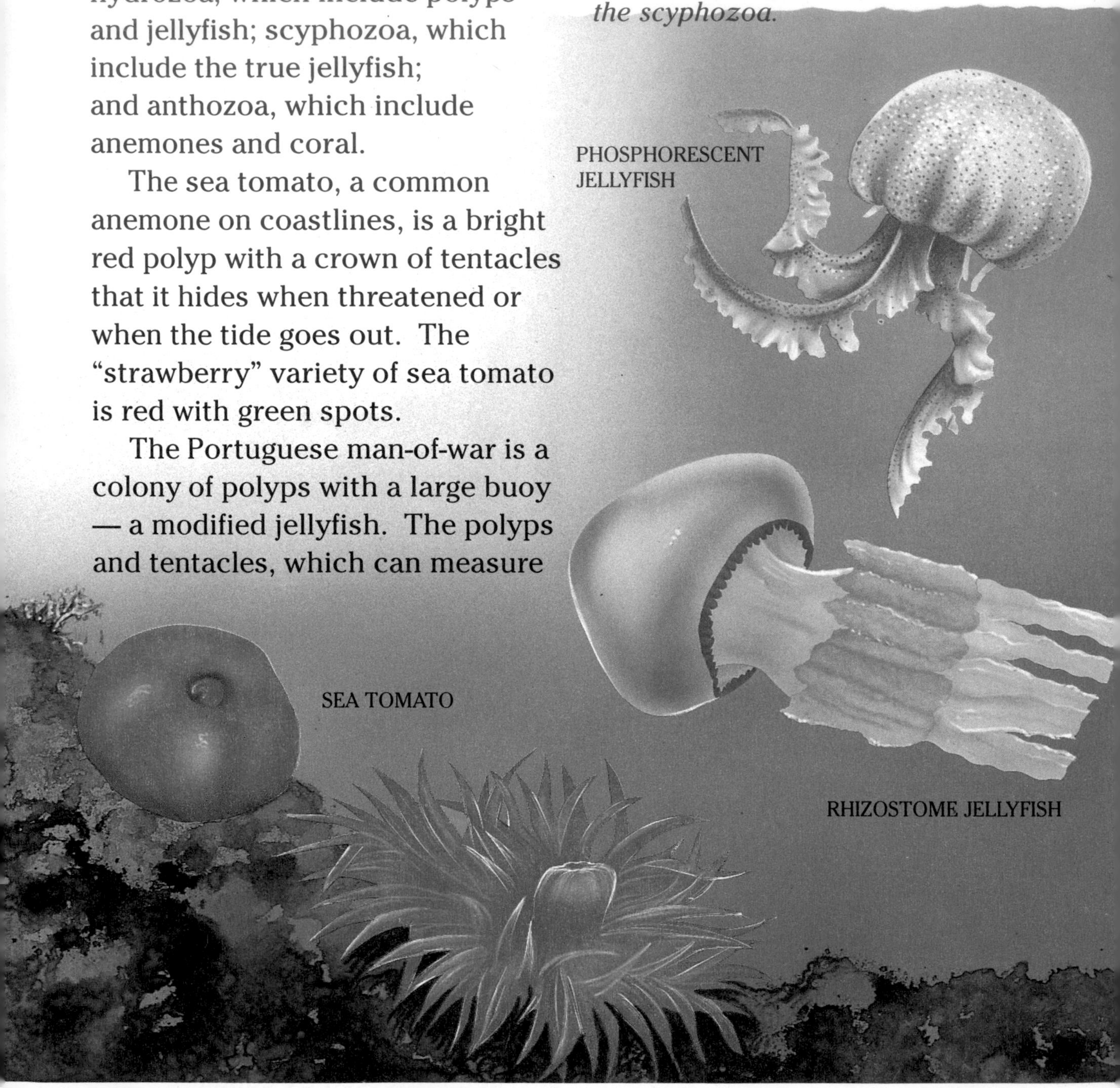

PHOSPHORESCENT JELLYFISH

SEA TOMATO

RHIZOSTOME JELLYFISH

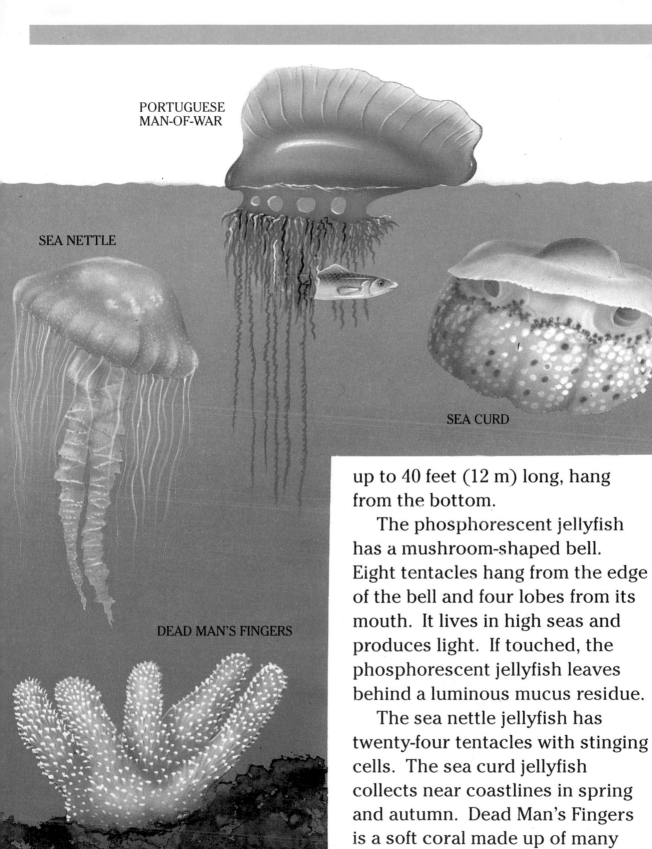

PORTUGUESE
MAN-OF-WAR

SEA NETTLE

SEA CURD

DEAD MAN'S FINGERS

up to 40 feet (12 m) long, hang
from the bottom.

The phosphorescent jellyfish
has a mushroom-shaped bell.
Eight tentacles hang from the edge
of the bell and four lobes from its
mouth. It lives in high seas and
produces light. If touched, the
phosphorescent jellyfish leaves
behind a luminous mucus residue.

The sea nettle jellyfish has
twenty-four tentacles with stinging
cells. The sea curd jellyfish
collects near coastlines in spring
and autumn. Dead Man's Fingers
is a soft coral made up of many
polyps connected by canals.

THE OBELIA JELLYFISH LIFE CYCLE

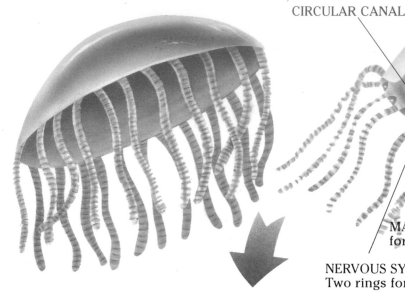

RADIAL CANAL

CIRCULAR CANAL

MANUBRIUM
forms gullet

GASTRIC CAVITY

FERTILIZED OVULE
When the female and the male jellyfish set their gametes free, they unite to form the ovule, which develops into the planula larva.

NERVOUS SYSTEM
Two rings form the jellyfish's nervous system; one is internal and the other is external.

OCELLUS
The ocelli, found at the base of the tentacles, can distinguish light but not shapes.

Jellyfish and polyps form part of the life cycle of many cnidarians. Polyps live anchored to the sea floor, alone or in colonies. Colonies are made up of various cnidarians with different missions. The simple internal anatomy of jellyfish and polyps is very similar. But the jellyfish's nervous system is more advanced. It also has gonads, or reproductive sexual organs, while polyps do not.

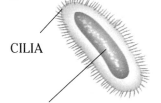

CILIA

PLANULA LARVA
This ciliated larva swims for a while, loses its cilia, then attaches itself to a surface. It secretes its protective coating and produces a colony of new individuals.

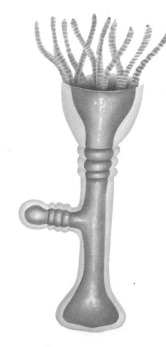

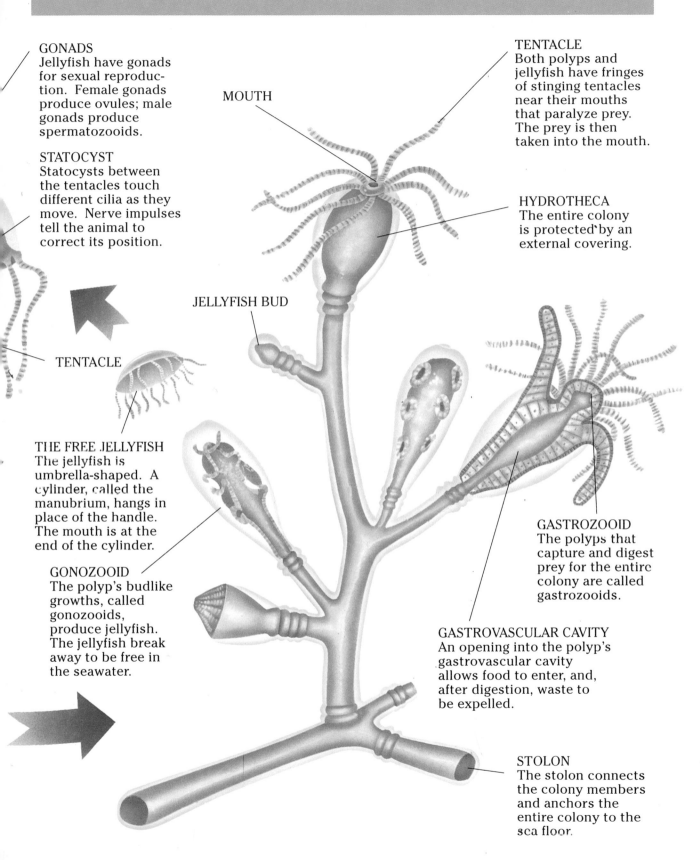

GONADS
Jellyfish have gonads for sexual reproduction. Female gonads produce ovules; male gonads produce spermatozooids.

STATOCYST
Statocysts between the tentacles touch different cilia as they move. Nerve impulses tell the animal to correct its position.

MOUTH

TENTACLE
Both polyps and jellyfish have fringes of stinging tentacles near their mouths that paralyze prey. The prey is then taken into the mouth.

HYDROTHECA
The entire colony is protected by an external covering.

JELLYFISH BUD

TENTACLE

THE FREE JELLYFISH
The jellyfish is umbrella-shaped. A cylinder, called the manubrium, hangs in place of the handle. The mouth is at the end of the cylinder.

GONOZOOID
The polyp's budlike growths, called gonozooids, produce jellyfish. The jellyfish break away to be free in the seawater.

GASTROZOOID
The polyps that capture and digest prey for the entire colony are called gastrozooids.

GASTROVASCULAR CAVITY
An opening into the polyp's gastrovascular cavity allows food to enter, and, after digestion, waste to be expelled.

STOLON
The stolon connects the colony members and anchors the entire colony to the sea floor.

JELLYFISH REPRODUCTION

The mamas and the papas

When the cnidarian goes into the jellyfish stage, it is responsible for sexual reproduction. Jellyfish are generally of separate sexes — males and females.

The jellyfish of the Chrysaora species is one of the few existing examples of an animal having both sexes. When they are born, the jellyfish are all males, but as they grow, they become female and can even be both sexes for a short time. The jellyfish releases ovules and spermatozooids into the sea, where fertilization takes place. It is also possible for fertilization to be internal. After fertilization,

When jellyfish do not go through the polyp stage, the planula develops directly into another jellyfish.

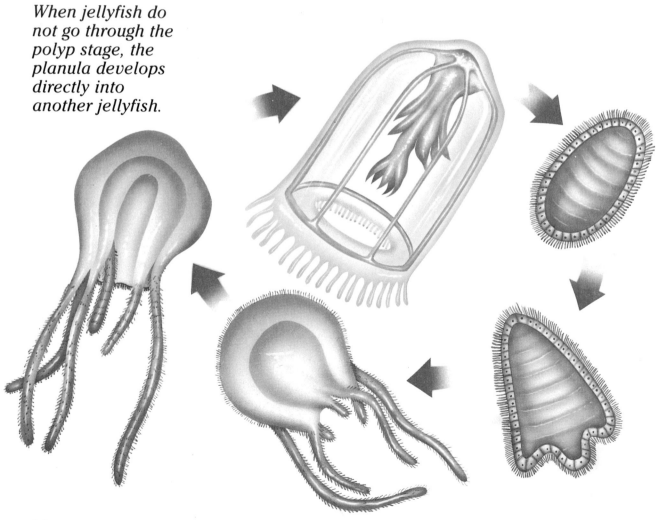

development can begin within the jellyfish itself.

A planula larva with cilia then forms and feeds on small organisms. Eventually, it anchors itself to a surface, where it turns into either a single polyp or a polyp colony.

In some species, the polyp stage does not exist. The planula develops tentacles and very quickly transforms into a jellyfish.

Polyps as well as some jellyfish can also reproduce through budding. They form new jellyfish from their bells or from a part near their mouths. Polyps can produce jellyfish and polyps by budding; jellyfish can produce only jellyfish when budding.

Both polyps and jellyfish, seen below, can reproduce through budding.

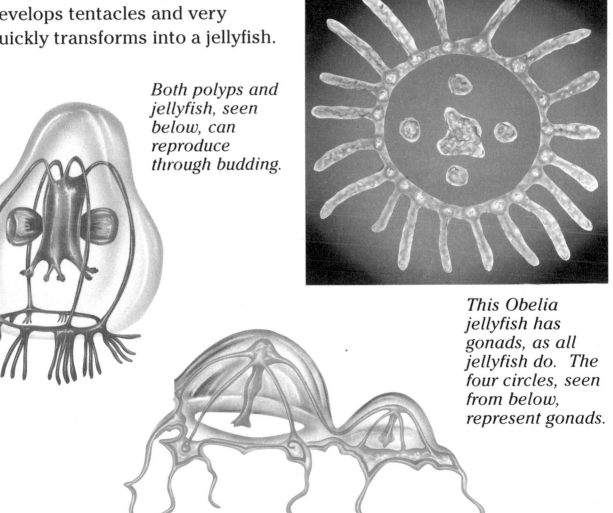

This Obelia jellyfish has gonads, as all jellyfish do. The four circles, seen from below, represent gonads.

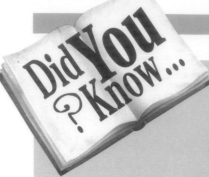

that some jellyfish know how to navigate?

By-the-wind sailor, or Velella, is a colony of floating polyps in the shape of a jellyfish. It drifts before the wind as it floats on the sea. The flat, gas-supported body carries a triangular sail, permanently slanted either to the right or to the left. This allows individuals to travel in different directions.

A Velella is 2 inches (5 cm) long and it descends as deep as 3,280 feet (1,000 m) to reproduce. After fertilization, the larva rises on drops of oil to the surface of the sea, where it becomes an adult.

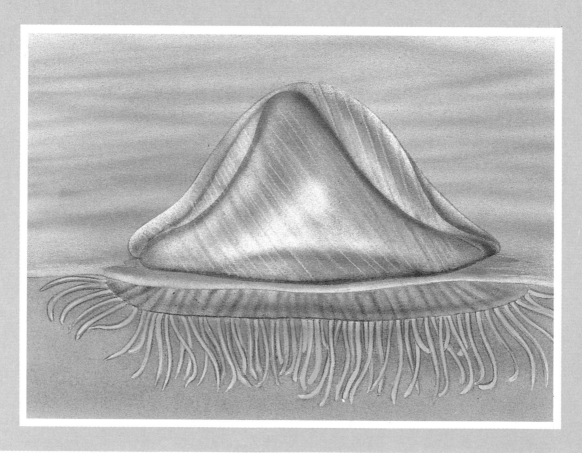

Jellyfish "prisoners"

Some species of jellyfish live attached to polyps in the colony. They spend their entire lives together. These jellyfish reproduce through the budding of one member of the colony. Ovules are fertilized inside the jellyfish. Eggs develop inside until they become planula larvae. When several larvae have developed, they are ejected from a transparent sac.

In other species, the planula continues its development inside the jellyfish and transforms into another larva that has tentacles. It then uses its tentacles to move away and start a new polyp colony.

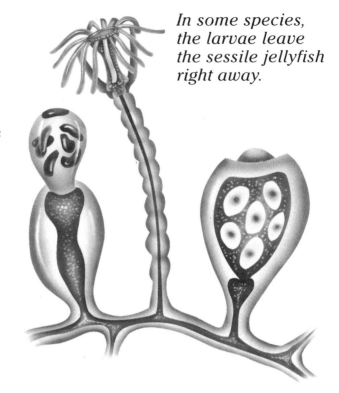

In some species, the larvae leave the sessile jellyfish right away.

Jellyfish grow among the tentacles of one of the polyps from its colony. Inside, it develops a planula, which becomes another larva that is freed to form a new colony.

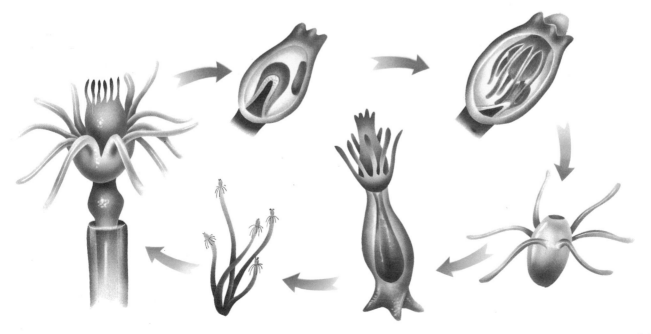

POLYP REPRODUCTION

Budding polyps

Polyps reproduce asexually through budding. They form buds that are somewhat similar to plant shoots. A new polyp is produced by each bud, but at certain times of the year, the buds form jellyfish.

All the members of a polyp colony share responsibility for work. Some individuals form jellyfish that can either be freed into the sea or remain attached to the colony. Some jellyfish pass through an unusual polyp stage that looks like a larva with tentacles. After several years of solitary life, this polyp divides into slices and forms stacks of young jellyfish that detach themselves at the right moment. Some take two years to become adult jellyfish. Others live only for a short time.

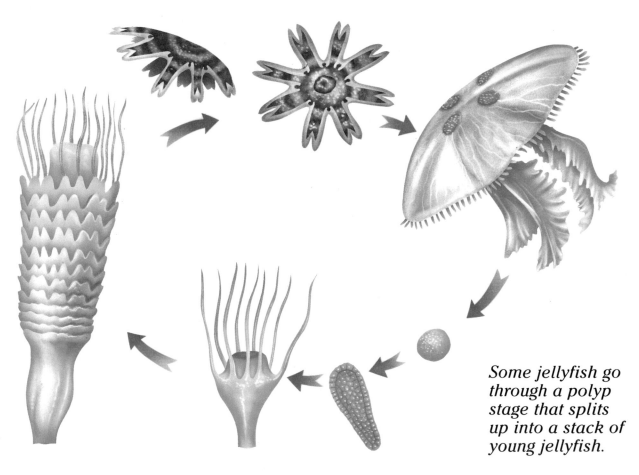

Some jellyfish go through a polyp stage that splits up into a stack of young jellyfish.

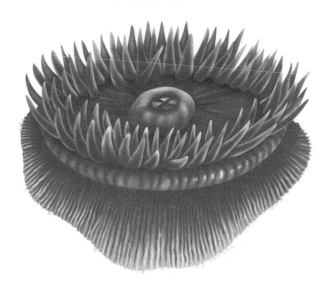

Anemones are polyps that never produce jellyfish. They can reproduce either sexually or asexually.

Anemones are solitary polyps that never produce jellyfish in their life cycle. They can reproduce asexually by splitting in two or through budding.

Sexual reproduction is also possible for the anemone. Cells are fertilized inside the female anemone or in the water.

Polyps are carnivores. They catch small crustaceans or fish with their tentacles, which contain small stinging organs like coiled harpoons, called nematocysts.

A polyp catches a crustacean with its tentacles. After swallowing, it once again waits for prey.

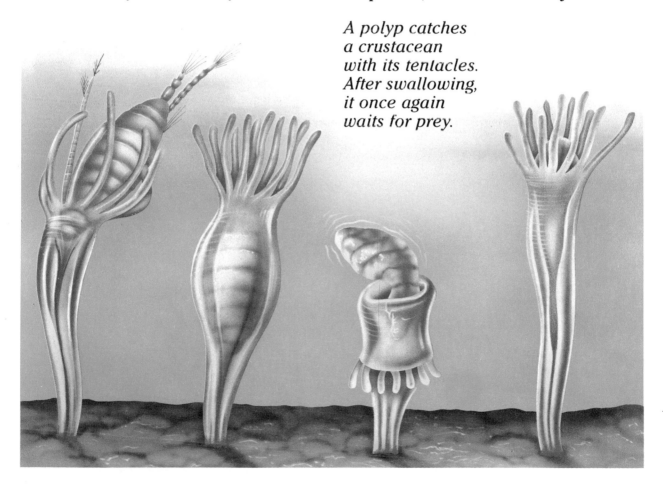

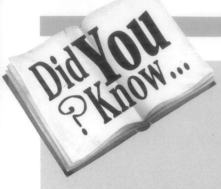

that there are "bodyguard" jellyfish?

Many animals of different species help each other on a regular basis. This relationship is called symbiosis.

A symbiotic relationship exists between some jellyfish and the small fish that follow them. Although these fish would make a good meal, the jellyfish do not eat them. From behind the tentacles, the fish can watch as their predators are hopelessly trapped.

In exchange for this protection, the jellyfish feed on the predators. The small fish then feast on the remains of the jellyfish's prey.

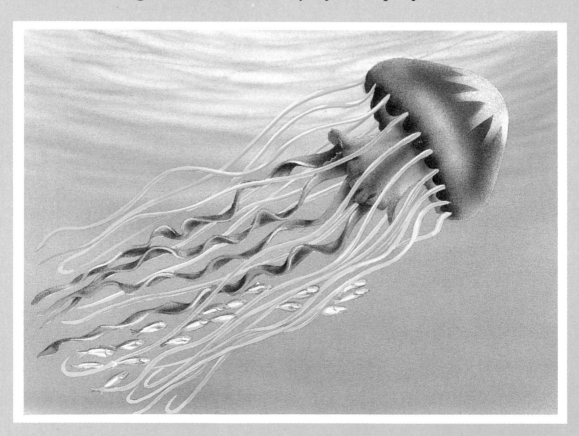

The unusual hydra

The freshwater hydra is a polyp that does not include jellyfish in its life cycle, yet reproduces in two ways. In summer, the polyp produces hydrae through budding. These hydrae become independent and attach themselves somewhere near their parent species.

Sexual reproduction takes place in autumn, when the hydrae develop either male or female gonads, or both at the same time. Spermatozooids released from the male gonads fertilize the ovules in the female gonads. The embryo that forms after fertilization can

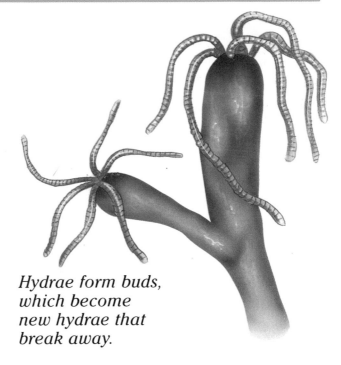

Hydrae form buds, which become new hydrae that break away.

detach itself from the body wall or cover itself with a protective coating for the winter.

Each ovary produces a bulging ovule. Spermatozooids fertilize the ovule, producing a daughter hydra.

THE ORIGIN OF THE JELLYFISH

In the old seas

Various animals have complex internal structures formed by tissues and organs that have specific functions. All animals have evolved from other more primitive and simple ones, very similar to the cnidarians.

Cnidarians, however, cannot be direct ancestors of other animals because they have stinging cells that no other animals possess: nematocysts. Jellyfish and polyps also have radial symmetry. This means they can be divided into several equal parts. Most animals can only be divided into two equal parts. Scientists have wondered

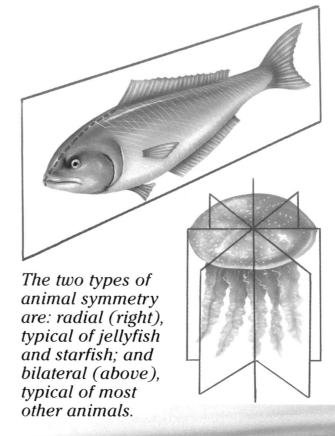

The two types of animal symmetry are: radial (right), typical of jellyfish and starfish; and bilateral (above), typical of most other animals.

Jellyfish have inhabited Earth's seas for a long time, alongside other creatures.

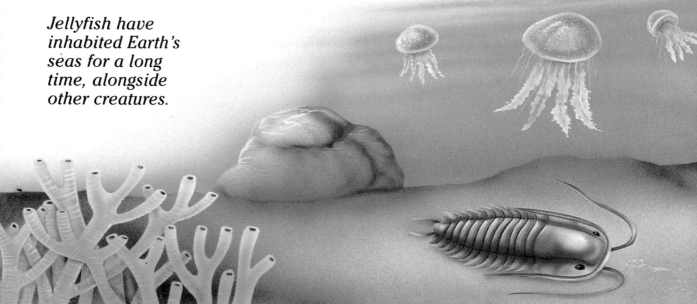

which came first in the cnidarians, the polyp or the jellyfish. It was most probably the polyp, since it can produce an individual through budding, with this individual having the ability to colonize a new area. The jellyfish might originate from an ancestral polyp's attempt to adapt itself to freer movement.

Six hundred million years ago, the seas were inhabited by primitive jellyfish. But because of their soft body matter, they have left no fossils, and little is known about them. Numerous remains of hard coral, however, do exist.

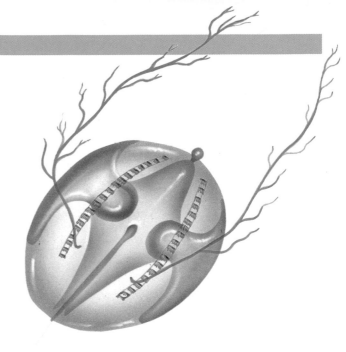

This animal, a Ctenophore, is a first cousin of the jellyfish. It shares similar characteristics, such as its very simple structure.

that some jellyfish live anchored to a rock?

Not all jellyfish swim freely in the sea. Some species live attached to rocks by means of an adhesive disk located on the side opposite the mouth. They live in protected coastal areas or on seaweed. The larvae of this jellyfish feed as a group. This way, they can catch prey that would be difficult to trap alone.

Cassiopeia jellyfish live face up on a sandy seabed and use their arms to catch prey.

Another species of jellyfish crawls on the seabed and attaches itself to vegetation with its tentacles.

JELLYFISH BEHAVIOR

Eat or be eaten

Jellyfish are carnivorous animals. Some of their prey are very large, but jellyfish can expand their bodies to eat the prey. Usually they trap prey without a chase. Everything edible within reach of their tentacles is immobilized by nematocysts and eaten.

Jellyfish enemies include some fish, such as the moon fish; and certain invertebrates, such as anemones or sea slugs. Jellyfish also make a good meal for leatherback turtles.

Jellyfish protect themselves from predators with their deadly

A sea slug devours a Portuguese man-of-war, one of the sea's most dangerous jellyfish.

harpoons. Sometimes, instead of swimming alone, they form groups, or banks, which increases their chances of defending themselves and escaping.

The leatherback turtle eats mostly jellyfish.

Dangerous umbrellas

Jellyfish have stinging tentacles that hang from the edges of their umbrella-shaped bodies. They usually have four tentacles, but this number increases with age.

Like polyps, jellyfish catch their prey using nematocysts, very effective weapons found on the tentacles, the mouth, and other body parts. Each nematocyst contains a capsule that holds a sort of harpoon wound up in a spiral. At the slightest contact with prey or predators, these harpoons are released to pierce the victim's skin, injecting a poisonous substance that paralyzes and usually kills it.

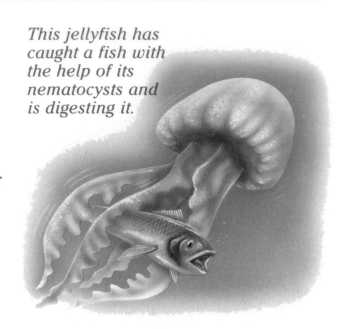

This jellyfish has caught a fish with the help of its nematocysts and is digesting it.

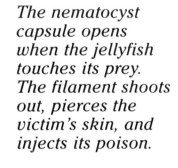

The nematocyst capsule opens when the jellyfish touches its prey. The filament shoots out, pierces the victim's skin, and injects its poison.

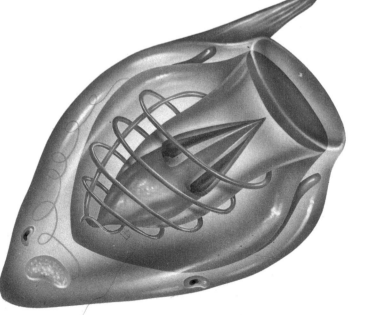

The stinging cells only work once, but new cells continually replace them in a few hours. Each nematocyst acts alone, but hundreds can attack unwary prey at the same time. Many jellyfish are harmless to humans, although a sting can be unpleasant. However, the sea wasp, a jellyfish that lives near coastal Australia, can kill a person.

Instead of piercing their prey, some nematocysts coil around body parts to secure them.

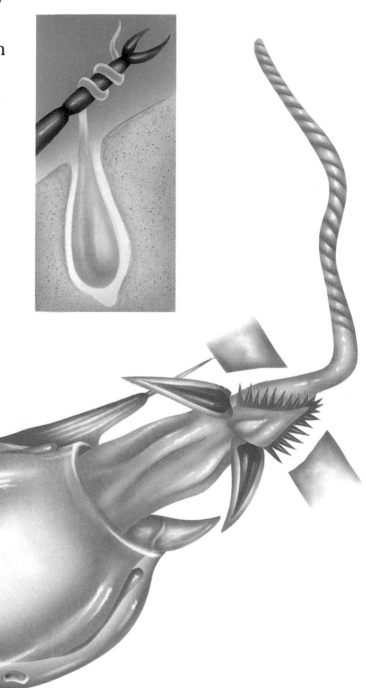

APPENDIX TO

SECRETS
OF THE
ANIMAL WORLD

JELLYFISH
Animals with a Deadly Touch

JELLYFISH SECRETS

▼ **Free rides.** Some anemones dwell in the shells of hermit crabs. The anemones protect

the crabs from enemies. The crabs provide transportation and food to the anemones.

▼ **The freshwater jellyfish.** One particular species of small jellyfish .4 inch (1 cm) in diameter lives in canals and rivers. Its polyp is smaller still, a difference so great that scientists once thought the polyps were from a completely different species. They were given the name *microhydra*.

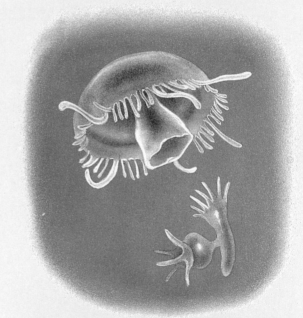

◄ **The clown fish's castle.** This bright fish lives among the anemone's tentacles. A mucus film prevents the anemone's poison from penetrating its skin.

Jellyfish with beacons. A multitude of luminescent jellyfish live in the coldest murky ocean waters. These jellyfish have thousands of dots of light that disappear and reappear in the darkness.

▶ **The narcomedusas.** At the polyp stage, these jellyfish live as parasites inside other jellyfish. They are lentil-shaped, and their tentacles rise above the bell instead of hanging below it.

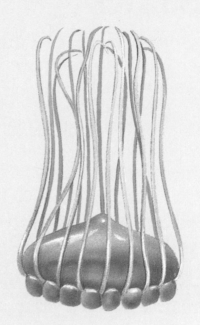

1. Jellyfish live:
a) in all the seas.
b) only on coasts where there are many sharks.
c) in all the seas, but also in rivers and lakes.

2. Asexual reproduction through budding is typical:
a) only of polyps.
b) only of jellyfish.
c) of both polyps and some jellyfish.

3. The jellyfish's main enemies are:
a) leatherback turtles and sea slugs.
b) carnivorous polyps.
c) sharks.

4. Polyps are reproduced:
a) from a jellyfish through budding.
b) from another polyp or through budding.
c) from a nematocyst.

5. Nematocysts contain:
a) drops of oil.
b) microscopic algae.
c) a filament coiled into a spiral.

6. Most of the jellyfish's nematocysts are located:
a) on their tentacles.
b) on their mouths.
c) on their bells.

The answers to JELLYFISH SECRETS questions are on page 32.

GLOSSARY

algae: a group of plants that grow in water. Algae do not have roots, stems, or leaves.

anemones: invertebrate animals that live in the ocean. Sea anemones usually attach themselves to rocks and shells and feed on plankton that they catch with their tentacles.

asexual: without male and female elements; asexual reproduction includes cell division, spore formation, and budding.

bell: the umbrella-shaped body of a jellyfish.

buoy: an object that floats in water.

carnivores: meat-eating animals.

cilia: short hair-like structures on a one-celled animal that help it move by beating rapidly.

coral: a type of polyp that lives in large colonies and secretes a hard substance for protection.

crustaceans: creatures with segmented bodies and an exoskeleton, or shell. Lobsters, shrimp, and crabs are crustaceans.

current: a flowing mass of water.

cylinder: an object that is shaped like a pipe or tube.

detach: to break away from; to break loose from something else.

edible: safe to eat; capable of being eaten.

embryo: an animal in the very earliest stages of growth, usually in an egg or its mother's uterus, after it has been conceived.

evolve: to change or develop gradually from one form to another over time to meet changing conditions.

fertilize: to make ready for reproduction, growth, or development.

gamete: a mature male or female germ cell used in the reproductive process.

gonads: reproductive glands that produce gametes and include ovaries and testes.

habitat: the natural home of a plant or animal.

harpoon: a barbed spear or javelin.

larva: in the life cycle of insects, amphibians, fish, and some other organisms, the stage that comes after the egg but before full development.

lethal: destructive; able to cause death.

lobe: a round projection or outgrowth.

luminous: shining; giving off light.

mucus: a slippery secretion that protects certain membranes.

navigate: to direct the course of a boat, plane, or some other craft.

nematocysts: stinging organs of cnidarians.

ovule: a small egg in an early stage of growth.

phylum: one of the main divisions of the animal kingdom. Jellyfish belong to the phylum Cnidaria.

polyp: an animal, such as a sea anemone, with a hollow body and tentacles.

predators: animals that kill and eat other animals.

prey: animals that are hunted, captured, and killed for food by other animals.

remains: what is left after something has died.

reproduce: to mate, create offspring, and bear young.

seaweed: a type of algae that lives in the sea.

sessile: permanently fixed or attached; attached directly to a base of some sort by a stalk or stem.

species: animals or plants that are closely related and often similar in behavior and appearance. Members of the same species are capable of breeding together.

statocyst: an organ of equilibrium occurring in some animals, including jellyfish, consisting of a fluid-filled sac containing a particle and a special cilium. The movement of the particle against the cilium produces a nerve impulse that causes the jellyfish to correct its position.

stolon: a root-like extension of the body wall that develops buds for new organisms that grow and remain attached.

symbiosis: the relationship of two or more different organisms that live in close association. A symbiotic relationship is one that provides benefits to each of the organisms.

symmetry: an exact matching of parts or sections on opposite sides of a dividing line or around a central point.

tentacles: narrow, flexible parts or limbs that certain animals use for moving around or catching prey.

tide: the regular rise and fall of the surface water levels of the oceans. The tides are caused by the interaction of the moon and the sun.

ACTIVITIES

◆ Visit an aquarium or an ocean world and observe the different kinds of jellyfish. In a notebook, make a simple sketch of each kind and write the name next to it. Do other fish and other sea life stay away from the jellyfish, or do they ignore them? What are the different colors of the jellyfish?

◆ On a map of the world, mark the areas where jellyfish that might sting humans occur. Shade these areas yellow and write in the name of the jellyfish. Now outline the areas of the world where jellyfish stings are dangerous to humans. Write in the name and color the areas red. Which kinds live in warm water and which in cooler waters? Is there a natural predator in each area that preys on the dangerous jellyfish?

◆ In areas with jellyfish that sting humans, what is being done to protect swimmers? Can you think of anything else that might protect bathers from being stung? What kind of treatment can be given to people who have been stung by jellyfish?

MORE BOOKS TO READ

Colors of the Sea (series). E. Ethan and M. Bearanger (Gareth Stevens)
Discover Ocean Life. Alice Jablonsky (Forest House)
Down in the Sea: The Jellyfish. Patricia Kite (A. Whitman)
The Illustrated World of Oceans. Susan Wells (Simon & Schuster)
Jellyfish. Lynn M. Stone (Rourke Corporation)
Rainbows in the Sea. Elizabeth R. Gowell (Watts)
Small Sea Creatures. Jason Cooper (Rourke Group)
Strange Eating Habits of Sea Creatures. Jean Sibbald (Silver Burdett)
Under the Sea. Kate Petty (Barron)

VIDEOS

Marine Biology: Life in the Tropical Sea. (Phoenix/BFA Films and Videos)
Marine Flowers. (International Film Bureau)
Marine Invertebrates. (Encyclopædia Britannica)
Simple Multicellular Animals: Sponges, Coelenterates, and Flatworms.
    (Benchmark Media)
Stinging-celled Animals: Coelenterates. (Encyclopædia Britannica)

PLACES TO VISIT

**Mystic Marinelife
    Aquarium**
55 Coogan Boulevard
Mystic, CT 06355

Aquarium du Quebec
1675 Avenue des Hotels
Sainte-Soy, PQ G1W 4S3

Taronga Zoo
Bradley's Head Road
Mosman, NSW
Australia 2088

**Vancouver Public
    Aquarium**
In Stanley Park
West Georgia Avenue
Vancouver, BC
V6B 3X8

**Sea World on the Gold
    Coast**
Sea World Drive Spit
Surfers Paradise
Queensland, Australia
4217

**Kelly Tarlton's
    Underwater World**
23 Tamaki Drive
Auckland, New Zealand

**Point Defiance Zoo and
    Aquarium**
5400 North Pearl Street
Tacoma, WA 98407

INDEX

Answers to JELLYFISH SECRETS questions:
1. c
2. c
3. a
4. b
5. c
6. a